THE PREPPER'S GUIDE TO THE DIGITAL AGE

SURVIVING THE RISKS OF DIGITAL
ARMAGEDDON AND NAVIGATING THE
PROMISE OF A TECHNOLOGICAL UTOPIA IN AN
UNCERTAIN FUTURE

SAM FURY

D1206896

WARNINGS AND DISCLAIMERS

The information in this publication is made public for reference only.

Neither the author, publisher, nor anyone else involved in the production of this publication is responsible for how the reader uses the information or the result of his/her actions.

CONTENTS

THANKS FOR YOUR PURCHASE

Get Your Next SF Nonfiction Book FREE!

Claim the book of your choice at:

www.SFNonfictionBooks.com/Free-Book

You will also be among the first to know of all the latest releases, discount offers, bonus content, and more.

Go to:

www.SFNonfictionBooks.com/Free-Book

Thanks again for your support.

INTRODUCTION

The Rapid Advancements of Technology and Their Effects on Society

Technology's rapid advancements have had a profound effect on society, revolutionizing how we live, work, and communicate. As prepper in the digital age, it is essential to comprehend these developments and their repercussions in order to remain prepared and resilient in an ever-evolving world. This chapter will examine the influence of technology across various aspects of daily life while outlining both its opportunities and challenges for preppers.

One of the greatest impacts of technological progress has been how we access and consume information. The internet and social media platforms have allowed people to stay informed globally, creating a free flow of ideas. For preppers this opens up an abundance of knowledge and resources that can assist with planning and preparation. Nevertheless, it also presents challenges such as having to distinguish between accurate and misleading sources, plus potential digital surveillance or data breaches.

Technological advancements have also transformed the way we work. Remote work and telecommuting are now commonplace, enabling people to work from anywhere with an internet connection. This shift has presented preppers with new opportunities for self-sufficiency; living off-the-grid while still participating in the digital economy. Unfortunately, it also leaves our digital infrastructure vulnerable to cyberattacks, necessitating preppers to invest in safeguarding their online presences.

Technology has had a major impact on society with the rise of automation and artificial intelligence (AI). Automation has brought about greater efficiency across various industries, from agriculture to manufacturing. Preppers now have access to cutting-edge tools and systems which can simplify their preparedness tasks. But as AI continues to replace jobs, it will become increasingly important for preppers to adapt by developing new skills and exploring alternative sources of income.

Technological progress has also enabled the rapid spread of the Internet of Things (IoT), in which everyday devices are linked to the internet and can communicate and share data. While this interconnection can be beneficial to preparedness by providing real-time info on weather patterns, natural disasters or potential threats; it also poses increased cybersecurity risks since hackers could potentially target these connected devices to access personal information or disrupt essential services.

Balancing Technological Benefits With Potential Risks

In today's digital age, prepper families must balance the numerous advantages of technology with potential risks associated with increased reliance on interconnected systems. While technology continues to progress at an astounding rate, it offers numerous opportunities for improved preparedness, communication, and self-sufficiency; however, potential threats such as cyber threats, data breaches, and infrastructure vulnerabilities must be carefully considered and managed for long-term resilience. This section will look at strategies for successfully striking this balance while making optimal use of technology in daily life as a prepper.

One of the most important components for prepping to balance technological benefits and risks is developing an extensive understanding of digital tools and platforms available. By staying abreast of technological advancements, preppers can make informed decisions about which systems and tools to adopt, as well as how to utilize them safely and efficiently.

Furthermore, having this understanding allows preppers to anticipate potential issues and create contingency plans in case these systems experience disruption or failure.

Privacy and security should be a top priority when taking advantage of technology. Strong passwords with two-factor authentication can protect online accounts from unauthorized access, while regularly updating devices and software helps mitigate potential vulnerabilities; using a virtual private network (VPN) protects data while maintaining privacy online; being cautious when sharing personal information on social media platforms reduces the risk of targeted attacks or identity theft.

Diversifying one's technology dependence is another effective strategy for managing risks. Preppers should avoid placing all their eggs in one basket by relying solely on one service provider or technology.

Having multiple communication channels like landlines, satellite phones and shortwave radios ensures preppers stay connected even during an internet outage or disruption. Furthermore, maintaining offline backups of important digital files and documents helps protect data against cyber threats or hardware malfunctioning.

Finally, prepper must cultivate a healthy skepticism when it comes to new technologies and the claims made by their proponents. Just because something is novel or innovative does not guarantee its security or dependability; so before incorporating new technologies into preparedness plans, prepper should critically evaluate them for vulnerabilities and long-term implications.

By taking advantage of technology while remaining aware of potential risks, preppers can successfully navigate the digital age. By staying informed, prioritizing privacy and security, reducing reliance on it, and exercising critical thinking, preppers can leverage its potential to enhance their preparedness while minimizing vulnerabilities.

The Importance of Being Prepared in the Digital Age

Prepare yourself mentally, emotionally and technologically for success in today's globalized world.

The digital age presents both unprecedented opportunities and unique challenges for preppers. As our lives become increasingly interconnected and dependent on technology, the importance of preparedness in this new era cannot be overemphasized. By understanding the dynamics of the digital landscape and devising strategies to mitigate potential risks, preppers can thrive in an ever-evolving world.

One of the primary reasons why preparedness is essential in the digital age is our increasingly vulnerable critical infrastructure. As society becomes more reliant on technology, so does the potential for disruption caused by cyberattacks, hardware malfunctions or natural disasters.

Preppers must be ready to handle prolonged outages in essential services like electricity, communication networks and transportation networks. Crafting contingency plans with alternative solutions in case these disruptions happen is essential to maintaining resilience in an increasingly interconnected world.

The digital age has seen an exponential rise in the availability and exchange of information. While this wealth of data presents many advantages for prepper, it also poses challenges such as information overload and misinformation. To be prepared in today's digital world, one must possess the capacity to differentiate accurate from false data, along with safeguard their own digital privacy. Cultivating digital literacy and critical thinking skills is essential for successfully navigating online information in a complex landscape.

Moreover, technological progress has introduced many tools and systems that can greatly assist preppers in their pursuit of self-sufficiency and resilience. Unfortunately, this reliance on digital systems may also present unexpected vulnerabilities. To remain prepared in today's digital age, one must stay abreast of technological advancements, assess their potential risks, and incorporate them into one's preparedness plans with consideration and balance.

The significance of preparedness in the digital age cannot be overemphasized. As our world becomes more interconnected and dependent on technology, preppers must adjust their strategies and mindsets to take advantage of all that this new era presents. By addressing critical infrastructure vulnerabilities, digital literacy issues, privacy concerns, and responsible technology adoption - preppers will ensure they are well-equipped to tackle whatever comes their way with confidence and resilience.

PART I

UNDERSTANDING THE DIGITAL LANDSCAPE

THE INTERNET AND ITS INFRASTRUCTURE

The digital age has usher in an era of convenience, connectivity and innovation. The Internet, a global network connecting computers and other devices, has become an integral part of our daily lives. For preppers looking to thrive in this new era, it is essential to comprehend its underlying infrastructure as well as any potential vulnerabilities. This chapter will give a comprehensive overview of both, outlining key components, potential risks, and strategies for safeguarding your online presence.

Components of the Internet Infrastructure

The Internet is a vast and intricate system composed of many elements that work together to enable data exchange. These components include:

- Internet Service Providers (ISPs): ISPs are organizations that give users access to the global internet. They maintain a network infrastructure connecting users to this global platform and offer services like broadband or dial-up connections.
- Routers and Switches: Routers and switches are networking devices that manage data traffic between devices and networks. Routers direct packets of information towards their desired destinations via the most efficient path, while switches enable communication within a given network.
- Data Centers: These are vast facilities that house servers and other networking equipment. They store and process vast amounts of data, providing the computational power behind various online services like websites, email services, and social media platforms.
- Domain Name System (DNS): The DNS transforms human-readable domain names like www.example.com

into IP addresses computers can understand, making it essential for navigating the internet.

- Undersea Cables: A global network of undersea cables transports most international data traffic. These cables span thousands of miles across oceans to link continents, enabling fast and dependable communication between nations.

Potential Risks and Vulnerabilities

The Internet's infrastructure is vulnerable to numerous risks and vulnerabilities, such as:

- Natural Disasters: Earthquakes, hurricanes, floods and other natural calamities can damage critical infrastructure components like undersea cables and data centers, disrupting Internet connectivity.
- Cyberattacks: Hackers can attack ISPs, data centers and the DNS to cause widespread outages and expose user information. Distributed Denial of Service (DDoS) attacks also take place which overwhelm targeted systems with an influx of traffic and make them inaccessible.
- Hardware Failure: Networking equipment such as routers and switches may experience hardware malfunction, leading to outages in internet services and disruptions in operations.
- Government Reactions: Governments can sometimes impose restrictions or block Internet access altogether during times of civil unrest or political unrest.

Prepping Strategies for a Resilient Digital Presence

Before embarking on any digital venture, it is important to have an idea of how you will protect your content.

As a prepper, it's essential to be ready for potential disruptions in your digital life. Here are some strategies for maintaining an resilient online presence:

- Diversify Your Internet Access: Don't rely solely on one ISP. If possible, have multiple connections from different providers so that if one goes down, you still have access to the web.
- Utilize a VPN: Virtual Private Networks (VPNs) encrypt your internet traffic, shielding it from hackers and maintaining privacy. In some cases, VPNs may even enable bypass of government-imposed restrictions and censorship.
- Maintain Offline Backups: Regularly back up essential digital files such as documents, photos, and videos to offline storage solutions like external hard drives or USB flash drives.
- Harden Your Devices: Make sure your devices are up to date with the latest security patches, use strong passwords, and enable two-factor authentication to reduce the risk of cyberattacks.
- Develop Alternative Communication Methods: In case of Internet disruptions, have alternative means of communication such as shortwave radio.

SOCIAL MEDIA AND ITS INFLUENCE ON SOCIETY

The rapid rise of social media has revolutionized how people communicate, receive information, and engage with one another. While its advantages are many, its downsides also exist. This chapter will examine both the positive effects of social media on society while also looking at ways preppers can utilize its potential while mitigating its risks.

The Social Media Landscape

Social media refers to an array of online platforms that permit users to create, share, and engage in social networking. Popular examples include Facebook, Twitter, Instagram, YouTube and TikTok. With the growth of these platforms has come major changes to how people connect, access information and form opinions.

Communication's Impact

Social media has revolutionized how we communicate with one another. It allows for rapid dissemination of information, allowing users to connect with friends, family, and acquaintances around the world. While this interconnectedness has its advantages - such as building virtual communities and increasing awareness about important issues - it may also foster misinformation or create echo chambers where individuals only interact with like-minded individuals who reinforce their beliefs.

Preppers can benefit from social media as a powerful tool for sharing ideas, resources, and information on preparedness. However, it is essential to verify the credibility of sources in order to avoid falling into echo chambers.

Privacy and Security Issues

Social media's widespread usage has raised serious privacy and security issues. Users often share personal information, location data, and other sensitive data on their profiles which puts them at risk of identity theft, cyberstalking, and other forms of cybercrime.

As a prepper, it is imperative to prioritize privacy and security when using social media platforms. This can be accomplished by setting privacy settings appropriately, being cautious with what information is shared, and using tools like VPNs for online identity protection.

Social Media and Mental Health

Social media plays a significant role in how we experience mental health today, both positively and negatively.

There is mounting evidence that excessive social media use can have detrimental effects on mental health. Comparing oneself to others online, cyberbullying, and social isolation are some of the factors contributing to anxiety, depression, and low self-esteem.

Preppers must be mindful of their social media consumption and take breaks when needed. Furthermore, it's essential to recognize the significance of striking a balance between online and offline connections.

Harnessing the Power of Social Media for Preparedness

Social media has become a great resource to prepare our communities. We must find ways to leverage its power for preparedness purposes.

Social media presents its own set of challenges, but when used responsibly can be an invaluable asset for preppers. Here are some ways to use social media effectively in preparation:

- Networking: Get together with other preppers and survivalists to share ideas, resources, and information. Doing this helps you stay abreast of the newest developments within the prepping community.
- Learning: Connect with reliable sources and experts in various fields related to preparedness, such as emergency medicine, gardening, and self-defense.
- Crisis Communication: Social media can be an invaluable resource for getting real-time updates during emergencies and disasters. Monitor relevant hashtags and follow official government accounts for accurate updates.
- Sharing Knowledge: Utilize social media platforms to share your expertise, experiences and tips with other preppers, adding to the collective knowledge of our community.

Social media has a profound impact on society, shaping the way we communicate, access information, and form opinions. Preppers must recognize both the advantages and potential hazards associated with its use. By taking necessary precautions and using social media responsibly, you can make the most of these platforms to improve preparedness levels, build connections, and stay informed.

As we navigate the digital age, it is essential for preppers to adjust and evolve with it. By understanding social media's impact on society and applying the strategies outlined in this chapter, you can continue growing and flourish in an ever-evolving world. Just remember to maintain a balance between online and offline lives, prioritize privacy and security, and always verify information's credibility before acting upon it. With these principles in mind, social media can be an incredibly helpful tool in your journey towards self-reliance and preparedness.

THE INTERNET OF THINGS (IOT) AND THE RISE OF SMART DEVICES

The Internet of Things (IoT) has become an essential element in today's digital age, connecting everyday objects to the internet and allowing them to collect, transmit, and process data. The proliferation of smart devices has had a profound effect on many aspects of modern life - from personal convenience to industry efficiency. In this chapter we'll look into what this means for preppers as well as examine some potential risks and rewards associated with IoT growth.

Comprehending the Internet of Things

The Internet of Things is the networked interconnection of physical objects, known as "smart devices," through the internet. These embedded electronics feature sensors, software and network connectivity which enables them to collect and exchange data with other systems or devices. Examples include smart thermostats, home security systems, wearables and appliances that utilize IoT capabilities.

Potential Rewards of Prepping

The rise of IoT devices offers preppers a wealth of opportunities to boost their preparedness and self-sufficiency. Potential advantages could include:

- Home Automation: Smart home systems can offer increased security, energy efficiency and convenience by automating tasks such as lighting control, temperature control and appliance management.
- Remote Monitoring: IoT devices enable preppers to keep an eye on their properties, gardens and livestock even when they're away from home. This is especially helpful for those with off-grid retreats or remote bug-out locations.

- Resource Management: Smart devices can assist preppers in monitoring and conserving resources like water and energy, helping them optimize their use and reduce waste.
- Early Warning Systems: Internet of Things sensors can detect and alert for various environmental hazards, like flooding, wildfires or severe weather events, giving preppers the opportunity to take proactive measures in order to safeguard their homes and families.

Potential Risques and Obstacles

Though IoT devices offer many advantages to preppers, there are also potential risks and challenges that must be considered:

- Security Vulnerabilities: The interconnectivity of IoT devices makes them vulnerable to hacking and cyberattacks. A compromised device could expose sensitive information, grant unauthorized access to your home, or even cause physical damage.
- Privacy Concerns: The vast amount of data collected by IoT devices can pose privacy risks if not properly secured. Preppers must exercise caution when sharing information with these devices and guarantee their data is safeguarded.
- Dependence on Connectivity: IoT devices typically rely on internet connectivity to function, which could be an issue during disasters or power outages. Preppers must have backup plans in place for such circumstances and not solely rely on IoT devices for their preparedness needs.
- Obsolescence: Rapid advances in technology may render IoT devices obsolete, necessitating regular updates or replacements to remain functional. Preppers must weigh the long-term viability and costs associated with incorporating IoT devices into their preparedness strategies.

Strategies for Integrating IoT Devices Responsibly

To maximize the potential of IoT devices while avoiding potential hazards, prepper should implement these strategies:

- Assess Your Needs: Determine which IoT devices best meet your preparedness objectives and provide practical advantages in your unique circumstance.
- Prioritize Security: Invest in devices with built-in security features and regularly upgrade their software to protect against potential cyber threats. Furthermore, utilize strong passwords and firewalls to bolster the security of your home network.
- Maintain Balance Dependency: While IoT devices can be invaluable tools for preparedness, do not become overly dependent on them. Make sure you have backup plans and non-digital solutions in place for essential tasks and functions.
- Educate Yourself: Stay abreast of the newest IoT technology developments, security vulnerabilities, and best practices for protecting privacy and data. By understanding this landscape thoroughly, you can make informed decisions about incorporating smart devices into your preparedness plan.

IoT devices and smart technology offer preppers both opportunities and challenges in the digital age. By carefully considering potential benefits, risks, and strategies for integrating IoT devices into your preparedness plan, you can improve self-sufficiency, security, resource management without compromising privacy or safety. As with any technology it is essential to stay informed and adaptable - finding a balance between taking advantage of IoT advantages while maintaining traditional preparedness skills and practices.

THE DARK SIDE OF THE DIGITAL AGE: CYBERCRIME, CYBERWARFARE, AND SURVEILLANCE

The digital age has enabled unprecedented connectivity and access to information, but it also opens the door for malicious activity and security risks. In this chapter, we'll delve into some of the darker sides of digital life, exploring topics like cybercrime, cyberwarfare, surveillance - as well as how preppers can prepare for and mitigate these risks.

Cybercrime

Cybercrime refers to any illegal activities conducted online. Common forms of cybercrime include:

- Identity theft: The unlawful acquisition and use of personal information to commit fraud or other illegal acts.
- Hacking: Unauthorized access to computer systems or networks with the aim to steal, modify, or destroy data.
- Phishing: Scammers attempt to obtain sensitive information such as passwords or credit card numbers by impersonating a reliable entity.
- Ransomware: Malicious software that encrypts files and demands payment in exchange for their release.

Preppers must take steps to protect their digital assets and personal information from cybercriminals. Enforcing strong passwords, updating software regularly, and exercising caution online can all help reduce the likelihood of being targeted by cybercriminals.

Cyberwarfare

Cyberwarfare refers to the use of digital attacks by nation-states or other organized groups to disrupt, damage, or gain unauthorized access to computer systems and networks of their adversaries.

Cyberattacks can target critical infrastructure like power grids, transportation systems, or communication networks with potentially disastrous results.

Preppers must be aware of the potential for cyberwarfare and its effect on their systems. Crafting contingency plans for disruptions in essential services like power or internet access can help preppers maintain resilience against cyber threats.

Surveillance

Surveillance has become more pervasive in the digital age. Governments, corporations and other entities can monitor and collect vast amounts of data on individuals through various methods such as internet browsing habits, social media activity and location tracking.

Preppers who value their privacy and autonomy must be aware of the potential for surveillance, taking steps to protect their personal information and digital footprint. Strategies such as using encrypted communication tools, virtual private networks (VPNs), and adopting privacy-conscious habits when browsing the internet should all be taken into consideration.

Preparing and Mitigating the Risks

To successfully prepare and mitigate the risks associated with cyber-crime, cyberwarfare, and surveillance, prepper should consider adopting the following strategies:

- Cyber Hygiene: Regularly update software, use strong passwords and multi-factor authentication, and be cautious online in order to reduce your exposure to cyber attacks.
- Data Backup: Regularly back up important files and documents onto external storage devices or secure cloud services in order to protect your information against loss or theft.

- Create Contingency Plans: Make plans for dealing with disruptions to essential services, like power or internet access, in the event of a cyber attack on critical infrastructure.
- Educate Yourself: Stay abreast of the latest cybersecurity, privacy protection, and potential threats to better comprehend their risks and take proactive measures to safeguard yourself.
- Adopt Offline Solutions: While digital tools and technologies offer many advantages, preppers should also build a strong foundation of traditional, offline preparedness skills and solutions in order to remain self-sufficient and resilient in the face of digital threats.

The digital age presents significant risks and challenges for preppers. By understanding the threats posed by cybercrime, cyberwarfare, and surveillance, and taking appropriate measures to safeguard your digital assets and personal information, you can maintain resilience and self-sufficiency in an increasingly interconnected world. It is essential to find a balance between taking advantage of digital technology and maintaining your privacy, autonomy and security. By staying informed, practicing good cyber hygiene, and creating contingency plans for potential disruptions, prepper can confidently and preparedly navigate the digital age with assurance, knowing they are equipped to tackle whatever challenges the future may bring.

PART II

ETHICAL AI DEVELOPMENT AND REGULATION

GOVERNMENTS AND ORGANIZATIONS: WHAT'S THEIR ROLE?

In today's digital world, governments and organizations play a pivotal role in shaping technology, cybersecurity, and emergency preparedness. As preppers endeavor to understand how these entities affect various aspects of preparedness - from privacy to information dissemination and essential services provision - it is essential that they be understood. In this chapter we will investigate this role of governments and organizations in today's digital landscape and consider how preppers can engage with these entities for greater preparedness and self-sufficiency.

Governments and Cybersecurity: What Is the Connection?

Governments are at the forefront of cybersecurity, working to safeguard critical infrastructure, combat cyber threats, and ensure citizens' privacy and security. Key components of government involvement in cybersecurity include:

- Cybersecurity Legislation: Governments create and enforce laws and regulations to safeguard against cybercrime, protect data privacy, and encourage responsible technology use.
- Cyber Defense: Governments maintain specialized agencies and teams dedicated to guarding against cyber threats and responding to attacks at a national scale.
- Public Awareness and Education: Governments often invest in public awareness campaigns and educational initiatives to spread cybersecurity best practices and create a culture of digital safety.

It is essential for preppers to stay informed about government regulations and guidelines related to cybersecurity and data privacy, since these can influence how they utilize digital tools and resources.

Emergency Preparedness and Response

Governments and organizations play a vital role in emergency preparedness and response, from natural disasters to pandemics to man-made crises. Their duties include:

- Emergency Planning: Governments create and implement emergency response plans to organize the efforts of various agencies and organizations during a crisis.
- Resource Management: Governments are accountable for managing and distributing essential resources such as food, water, and medical supplies during emergencies.
- Communication and Information Sharing: Governments and organizations are essential sources of emergency information, providing updates and guidance through various channels including digital platforms.

Preppers can benefit from accessing government and organizational resources related to emergency preparedness, such as disaster response plans, educational materials, and alerts. These tools allow prepper to enhance their own plans for preparedness while staying informed during emergencies.

How can you engage with governments and organizations?

To maximize the advantages of government and organization resources and expertise, preppers should consider adopting these strategies:

- Stay Informed: Make it a habit of regularly monitoring government and organizational websites, social media accounts, and other communication channels for updates on cybersecurity threats, emergency preparedness measures, and other pertinent topics.
- Participate in Training and Educational Opportunities: Take advantage of government and organizational training

programs, workshops, and webinars to develop your
preparedness skills and knowledge.

- Collaborate With Local Community: Build relationships
  and collaborate on preparedness initiatives with local
  government officials, first responders, and community
  organizations.
- Advocate for Responsible Policies: Stay informed on
  proposed legislation and policies related to privacy, security,
  and emergency preparedness. Promote responsible
  measures that safeguard individual rights while building
  resilience.

Governments and organizations play a vital role in the digital age,
directly impacting preparedness. By understanding their responsibil-
ities and resources, as well as actively engaging with them, preppers
can improve their own preparedness while contributing to creating a
more resilient and secure digital world. Successfully navigating this
digital era as a prepper requires striking an equilibrium between
self-reliance and collaboration, along with an intimate grasp of technology's ever-evolving interplay between privacy and security.

INVOLVEMENT OF THE AI COMMUNITY

Recent advances in artificial intelligence (AI) and machine learning technologies have drastically transformed many aspects of modern life, from communication to healthcare and entertainment. Within preparedness and self-sufficiency planning, the AI community plays a pivotal role in shaping technology's future applications for emergency situations. In this chapter, we'll investigate how the AI community fits into this digital age, outlining potential benefits, challenges, and ethical considerations preppers should be aware of when utilizing AI-driven technologies.

AI Applications in Preparedness and Emergency Response

Artificial intelligence has many applications which can assist emergency managers in planning ahead for possible crises or situations that may arise.

The AI community has made significant advances in creating tools and technologies that can assist emergency preparedness and response efforts. Some key applications of these advancements include:

- Early Warning Systems: AI-powered algorithms can glean vast amounts of data from various sources, like weather sensors and social media feeds, in order to predict natural disasters, disease outbreaks, and other emergencies with extreme accuracy.
- Resource Allocation and Logistics: AI technologies can optimize the distribution of essential resources such as food, water, and medical supplies during emergencies by taking into account factors like supply availability, transportation networks, and affected population needs.
- Search and Rescue Operations: AI-powered drones and robots can assist in search and rescue operations, traversing difficult terrain to locate survivors and deliver aid.

- Decision Support Systems: AI-powered tools can assist emergency responders and preppers in making more informed decisions by analyzing complex data sets, forecasting potential outcomes, and providing actionable insights in real-time.

Challenges and Ethical Considerations

As AI technologies become more deeply embedded into emergency preparedness and response activities, several challenges and ethical considerations must be addressed, such as:

- Data Privacy and Security: The collection and analysis of large amounts of data by AI systems may raise concerns about personal privacy and data security, particularly if sensitive information is compromised or misused.
- Bias and Discrimination: AI algorithms can reinforce existing biases if they are trained on inaccurate data or not designed with fairness and equity in mind.
- Reliability and Accountability: AI systems have the potential to make mistakes or produce unexpected outcomes, raising questions about their dependability as well as the accountability of their developers and users in emergency situations.
- Job Displacement: The increasing automation of tasks through AI technologies could potentially result in job displacement in certain sectors, such as emergency response and disaster management, potentially impacting the livelihoods of professionals employed there.

Connecting With the AI Community

To effectively navigate the rapidly advancing world of AI-driven technologies in preparedness and emergency response, preppers should consider adopting these strategies:

- Stay Informed: Regularly monitor AI research, news and developments to gain a better insight into the potential applications, limitations and ethical considerations related to AI in emergency preparedness.
- Participate in AI-Related Events: Attend conferences, workshops and webinars focused on AI and its applications in preparedness and emergency response to learn from experts and join discussions about the future of AI in these fields.
- Develop AI Literacy: Gain a fundamental comprehension of AI concepts, technologies, and ethical considerations so you can make informed decisions about incorporating AI-driven tools into your preparedness efforts.
- Advocate for Ethical AI: Promote the development and use of AI technologies that prioritize privacy, security, fairness, and transparency; support policies and initiatives that promote ethical AI practices in emergency preparedness and response.

The rise of artificial intelligence (AI) in today's digital world presents both opportunities and challenges for preppers alike. By staying informed about AI advances, engaging with the AI community, and advocating for ethical AI practices, prepper can reap the potential advantages of these cutting-edge technologies while mitigating any associated risks or ethical dilemmas. As AI continues to revolutionize emergency preparedness and response, preppers must remain adaptable and proactive, taking advantage of the potentialities offered by these technologies while being mindful of potential pitfalls they could present. By having a comprehensive understanding of AI's role in the digital age and its implications for preparedness, preppers will be better equipped to navigate the uncertainties that come with an increasingly interconnected world, maintaining their resilience and self-sufficiency against emerging threats and obstacles.

PART III

THE PROMISE OF A TECHNOLOGICAL UTOPIA

ARTIFICIAL INTELLIGENCE: SHAPING THE FUTURE

As we move further into the digital age, artificial intelligence (AI) continues to play an increasingly significant role in shaping our world. In this chapter, we'll investigate some of its wide-reaching implications and how they may impact preppers in various aspects of their lives - from communication and resource management to security protocols and beyond.

Communication

Communication is paramount for prepper success, as it allows the exchange of vital information and coordination during emergencies. AI-powered communication systems like natural language processing (NLP) and machine translation will break down language barriers to allow real-time conversation with people from various linguistic backgrounds. This proves invaluable in situations requiring international cooperation or when seeking refuge abroad.

Resource Management

AI has the potential to revolutionize resource management for preppers by optimizing supply allocation and other resource usage. AI algorithms can analyze resource consumption patterns such as water consumption or energy use and make predictions on future usage patterns. By recognizing inefficiencies and suggesting ways to improve resource utilization, AI can assist preppers in achieving greater self-sufficiency and sustainability.

Agriculture and Food Production

In order to sustainably increase agriculture and food production, there is a need for increased research.

AI can revolutionize agriculture and food production. Precision agriculture powered by AI maximizes crop yields while using less water, fertilizers, and pesticides - leading to greater food security with reduced reliance on external resources - an invaluable asset for preppers. Furthermore, AI is being utilized in vertical farming technologies and other innovative agricultural technologies that allow food production in confined spaces such as urban environments or underground bunkers.

Security and Surveillance

Security is a top concern for preppers, and AI has the potential to significantly enhance safety measures. AI-powered surveillance systems can process vast amounts of data from cameras, sensors and other sources to detect potential threats and unusual activities. Once alerted, preppers can take appropriate action such as deploying countermeasures or evacuating the area. Furthermore, AI can be employed in cybersecurity solutions by shielding digital assets from hackers and other malicious actors.

Medical Care

AI's impact on medical care can be a game-changer for emergency preparedness. AI-driven diagnostic tools and telemedicine platforms give preppers remote access to expert medical expertise even when traditional healthcare services are unavailable. Furthermore, AI-powered devices like wearable health monitors provide real-time information on vital signs and other health indicators, enabling timely interventions and improved overall health management.

Education and Skill Development

Education and skill development are integral parts of successful development for any person, both professionally or personally.

As the world becomes more complex, prepper must continuously learn and acquire new skills to stay prepared. AI can assist this process by offering personalized education and training programs. Machine learning algorithms can assess an individual's learning style, strengths, and weaknesses to customize content for maximum effectiveness. Doing so ensures that preppers acquire the necessary knowledge and abilities to cope more efficiently in the digital age.

The Dark Side of AI

While AI offers great promise in many aspects of prepping, it is essential to be aware of its potential risks as well. AI's rapid advancement may result in job displacement, social unrest and increased dependence on technology - creating new vulnerabilities for preppers. Furthermore, there is always the risk that AI could be weaponized or misused by malicious actors; so as preppers we must remain vigilant and develop strategies to mitigate these potential issues so as to stay ahead in today's digital age.

Artificial intelligence has the potential to have a major impact on our future, and its influence over prepping is no different. By harnessing the potential of AI, preppers can enhance their capabilities in fields such as communication, resource management, agriculture, security, medical care and education. However, it is essential to remain mindful of potential risks and difficulties that this technology may present. By staying informed and adaptable, prepper can take advantage of AI while mitigating its potential drawbacks, leaving them well-equipped to thrive in the digital age.

VIRTUAL AND AUGMENTED REALITY ARE REVOLUTIONIZING THE WAY WE INTERACT WITH THE WORLD

Virtual reality (VR) and augmented reality (AR) technologies have revolutionized how we perceive and engage with the world around us. This chapter will investigate how VR/AR technologies are impacting our daily lives, as well as how preppers can use them to boost their preparedness in the digital age.

Appreciating Virtual and Augmented Reality

Virtual reality (VR) is an immersive technology that creates a completely artificial environment, usually through the use of headsets and other sensory input devices. On the other hand, augmented reality (AR) overlays digital information onto the real world through smartphones or AR headsets. Both technologies have the potential to revolutionize various aspects of life such as entertainment, education, communication and emergency response.

Training and Skill Development

One of the most promising applications of VR and AR for preppers is in training and skill development. These technologies create realistic simulations of emergency scenarios, giving preppers a safe space to practice their response skills in an organized setting. Furthermore, VR and AR can be employed to teach complex topics like first aid or navigation in an engaging and immersive manner, making learning more efficient and enjoyable.

Remote Communication and Collaboration

In times of emergency, effective communication and collaboration are paramount to survival. Virtual reality (VR) and augmented reality (AR) have the potential to revolutionize remote communication by enabling preppers to work together even when separated by great distances. Through shared virtual environments, people can collaborate by exchanging information and working together toward solving problems or coordinating actions.

Improved Situational Awareness

Augmented reality (AR) can provide preppers with invaluable real-time data and insights about their environment, increasing situational awareness. For instance, AR can overlay information about nearby resources, hazards or points of interest directly onto a user's field of view; this is especially helpful in unfamiliar or rapidly changing environments so that emergency responders make better informed decisions and respond more efficiently during crises.

Disaster Response and Recovery

VR and AR can play an essential role in disaster response and recovery initiatives. These technologies enable detailed digital reconstructions of disaster-struck areas, allowing responders to assess damage, identify hazards, and plan recovery operations from afar. Moreover, AR helps guide emergency responders on the ground by providing real-time information and navigational aid so that they reach those in need as quickly and safely as possible.

Mental Health and Well-Being

It is important to recognize the psychological toll disasters and emergencies can have on individuals, and VR and AR offer invaluable tools for supporting mental health and well-being. These technologies create soothing virtual environments which help individuals cope with stress, anxiety or trauma more effectively. Furthermore, VR/AR technologies facilitate remote mental health services - providing preppers access to counseling from professionals even when traditional services may not be available.

Acknowledging the Benefits and Risks

VR and AR technologies offer numerous potential advantages for prepper, but it is essential to consider the risks and challenges associated with their use. Concerns such as privacy, security, and over-reliance on technology must be carefully assessed in order to maximize the advantages while minimizing any drawbacks that might occur.

Virtual and augmented reality technologies are revolutionizing how we engage with the world, offering preppers new opportunities as well as challenges. By taking advantage of VR/AR potentials, preppers can enhance their training, communication, situational awareness, disaster response capabilities, mental wellbeing - but it's essential to balance their benefits with potential risks or drawbacks that arise from using them. Staying informed and adaptable are essential for successfully utilizing these powerful tools while mitigating any potential drawbacks they may present. Doing so allows preppers to maximize the benefits from VR/AR technologies while mitigating potential drawbacks so they remain prepared to navigate the ever-evolving landscape of digital age.

RENEWABLE ENERGY AND SUSTAINABLE TECHNOLOGY HAVE MASSIVE POTENTIAL

As technology advances, the need for sustainable and reliable energy sources grows ever more pressing. Renewable energy and sustainable technologies offer the promise of a cleaner, more resilient future - essential for preppers seeking self-sufficiency and independence from traditional energy grids. This chapter will discuss these potentials of renewable sources and sustainable technologies and how they can be integrated into long-term preparedness plans.

Gaining An Understanding of Renewable Energy and Sustainable Technologies

This module provides an introduction to renewable energy sources and sustainable technologies, helping you better understand their effects on society.

Renewable energy comes from natural resources that replenish themselves periodically, such as sunlight, wind, and water. Sustainable technologies refer to innovative methods, tools or systems designed to minimize environmental impact while maximizing resource use efficiency. When combined together, renewable energy and sustainable technologies offer an eco-friendly and self-sufficient solution for meeting energy needs while decreasing reliance on conventional sources like fossil fuels.

Solar Energy

Solar energy is one of the most accessible and popular renewable sources for preppers. Photovoltaic (PV) solar panels convert sunlight into electricity, providing a clean source of power for homes, appliances, and electronic devices. Solar energy systems can be scaled to meet individual needs; they're especially well-suited to remote locations or off-grid living situations - making them an attractive option for preppers.

Wind Energy

Wind energy harnesses the energy of wind through turbines, converting kinetic energy into electricity. While more unpredictable than solar power, wind systems can still provide an effective supplement or alternative for those in regions with regular wind patterns. In rural areas, small-scale wind turbines may even be installed on private properties as an additional source of renewable energy for those seeking greater self-sufficiency.

Hydroelectric and Micro-Hydro Systems

Hydroelectric power relies on water's movement to generate electricity. While large-scale hydroelectric projects may not be feasible for individual preppers, micro-hydro systems offer an alternative when those with access to a reliable water source such as streams or rivers. These smaller systems generate steady amounts of electricity, making them attractive options for preppers in suitable places.

Sustainable Technologies for Energy Storage and Efficiency

Sustainable technologies offer opportunities to store energy more efficiently while increasing energy output.

Renewable energy sources aside, sustainable technologies are crucial in promoting energy efficiency and storage. Energy storage devices like batteries or fuel cells can store excess power generated by renewable systems to guarantee a constant supply of electricity even during periods of low production. Furthermore, energy-saving appliances and smart home systems help preppers reduce their reliance on conventional power sources by helping them minimize consumption.

Biogas and Biomass Energy Production

Biogas and biomass energy sources offer preppers the unique chance to utilize organic waste materials as sources of renewable energy. Biogas can be generated through anaerobic digestion, which converts organic material such as food scraps, animal manure, and agricultural waste into combustible gas. On the other hand, biomass energy involves burning organic material like wood or crop residues for heat or electricity production. Both of these energy sources add another layer of self-sufficiency and resourcefulness to preparation efforts.

Stressing the Importance of Energy Independence

Preppers recognize the importance of energy independence when ensuring long-term preparedness and resilience. Renewable sources and sustainable technologies offer a more reliable, self-sufficient, and environmentally friendly alternative to conventional power grids. By including these systems into their plans for preparedness, preppers can reduce their vulnerability to power outages, grid failures, and other energy-related disruptions.

In today's digital age, renewable energy sources and sustainable technologies offer preppers cleaner, self-sufficient solutions to meet their energy needs. By leveraging the potential of solar, wind, hydro-electric, biogas and biomass energy resources with energy storage and efficiency solutions, preppers can achieve greater energy independence and minimize their vulnerability to disruptions in conventional power sources. As we navigate the digital age, embracing renewable energy and sustainable technologies must be part of any comprehensive prepper's strategy. In the next chapter, we'll look at how preppers can build and sustain resilient communities in this digital era by encouraging cooperation, resource sharing, and mutual aid to guarantee everyone's wellbeing and survival.

TECHNOLOGY'S EFFECT ON HEALTHCARE, EDUCATION, AND COMMUNICATION

The digital age has brought about extraordinary advances in technology, with far-reaching effects across various aspects of our lives. In this chapter, we'll investigate how technology affects healthcare, education and communication - as well as how preppers can use these advances to boost their preparedness levels and overall wellbeing.

Healthcare

The digital age has revolutionized healthcare, offering preppers new tools and resources to manage their wellbeing and access medical services.

Telemedicine

Telemedicine allows patients to consult with healthcare professionals remotely through digital communication tools like video calls or chat applications. This can be especially helpful for preppers in remote locations or during emergencies when traditional healthcare services may not be accessible.

Wearable Health Monitors

Wearable devices that monitor vital signs and other health indicators can provide real-time insight into an individual's wellbeing, enabling early detection of potential issues and timely interventions. These devices are especially beneficial to preppers, helping them keep their health and well-being stable even in challenging conditions.

Electronic Health Records (EHRs)

EHRs enable healthcare providers to store and access patient information digitally, streamlining the process of providing medical care while improving coordination among professionals. Preppers will find having access to their EHRs essential in emergencies; having vital health data readily accessible when required.

Education

Technology has revolutionized the way we learn and acquire new skills, providing preppers with innovative methods to increase their knowledge base and preparedness.

Online Learning

Online platforms and e-learning resources have enabled preppers to access a wealth of information and educational content from anywhere. This can be especially beneficial for developing new skills or staying abreast of preparedness strategies.

Virtual and Augmented Reality

As discussed previously, virtual and augmented reality technologies offer immersive and engaging learning experiences, facilitating more effective skill development and training. Preppers can use these tools to practice emergency response scenarios or learn complex subjects more interactively.

Communication

The digital age has fundamentally transformed communication, providing preppers with new tools and opportunities to stay connected and share knowledge.

Social Media and Online Communities

Social media platforms and online forums give preppers the opportunity to connect with like-minded individuals, exchange ideas, and access invaluable resources. These communities can offer invaluable support during emergencies or when seeking advice on preparedness strategies.

Emergency Communication Systems

Digital communication tools, such as satellite phones, GPS devices and smartphone applications can significantly help preppers stay connected during emergencies. These devices facilitate response efforts, share essential information and guarantee the safety of individuals and communities alike.

Assessing the Benefits and Potential Risks

It is essential to balance both benefits and risks when making any major decision.

Technology has undoubtedly revolutionized healthcare, education, and communication - but it is essential for prepper to consider potential risks and vulnerabilities.

Matters such as privacy, cybersecurity, and over-reliance on technology must be carefully assessed in order to maximize their advantages while minimizing any negative repercussions.

Technology's impact on healthcare, education and communication has the potential to significantly boost preppers' readiness and overall well-being in today's digital world. By taking advantage of advances in telemedicine, wearable health monitors, online learning, virtual and augmented reality technology, as well as digital communication tools, preppers can better equip themselves to tackle the challenges of the modern world.

However, it is essential to weigh the advantages of technology with a careful consideration of its risks and challenges, enabling preppers to remain adaptable and resilient in an ever-evolving digital landscape.

PART IV

THE RISKS OF DIGITAL ARMAGEDDON

CYBERSECURITY THREATS AND DATA BREACHES

As we navigate the digital landscape of the 21st century, it is essential to be aware of the myriad cybersecurity threats and data breaches that could threaten our lives. This chapter will cover common cyberthreats and breaches, their causes, as well as practical steps you can take to safeguard yourself against their damaging effects.

Recognizing Cybersecurity Threats

Understanding cybersecurity threats is essential for any serious business owner.

Cybersecurity threats are attempts to compromise the confidentiality, integrity or availability of your digital assets such as personal information, financial data and sensitive documents. Common cybersecurity risks include:

- Phishing: When an attacker poses as a reliable entity and sends fraudulent emails or messages, they aim to deceive recipients into divulging sensitive information or installing malware.
- Ransomware: Malicious software that encrypts victim's files, making them inaccessible until a ransom payment is made - usually in cryptocurrency.
- Man-in-the-Middle (MITM) attacks: An attacker intercepts communication between two parties to steal sensitive information or manipulate the conversation.
- Distributed Denial of Service (DDoS) attacks: An attacker floods a network, website, or online service with an excessive amount of traffic, causing it to crash and become unavailable for legitimate users.
- Zero-day Exploits: Attackers take advantage of vulnerabilities in software and hardware before the developer has the chance to create and distribute a patch.

Recognizing Data Breaches

Data breaches occur when an unauthorized individual gains access to sensitive information within an organization, such as personal data, financial records or intellectual property. Data breaches can have far-reaching repercussions for organizations; from identity theft and financial loss to reputational damage.

Securing Your Digital Life

Securing Your digital life is of utmost importance for protecting it against threats and attacks.

To safeguard yourself against cybersecurity threats and data breaches, adhere to these best practices:

- Create strong, unique passwords: Combine letters, numbers and special characters to create complex passwords that cannot be broken. Avoid using the same password across multiple accounts.
- Enable two-factor authentication (2FA): 2FA adds an extra layer of protection by requiring you to provide a secondary method of verification, such as text message or fingerprint scan, in addition to your password.
- Regularly Update Software and Hardware: Ensure your operating systems, applications, and devices are up to date with the latest security patches to guard against known vulnerabilities.
- Install reliable antivirus and anti-malware software: Use trusted security software to detect, prevent, and eliminate malware and other threats.
- Be wary of phishing attempts: Be wary of unsolicited emails and messages that request personal information or require you to click a link or download an attachment.
- Use a VPN when connected to public Wi-Fi: A virtual private network (VPN) encrypts your data, making it

difficult for attackers to intercept and read your online activities.

- Secure Your Home Network: Update the default usernames and passwords on your router, enable WPA3 encryption, and make sure the firmware of your router is always up-to-date.
- Regularly Monitor Your Online Accounts: Make sure to regularly check your bank, credit card and other sensitive accounts for suspicious activity and report any discrepancies promptly.
- Backup Your Data: Regularly create and store copies of your important files and data on an external device or cloud service to guard against ransomware attacks and other data loss scenarios.
- Stay Informed: Stay abreast of the newest cybersecurity threats and trends to take a proactive approach to protecting your digital security.

By understanding and employing these security measures, you can reduce your vulnerability to cybersecurity threats and data breaches. With technology constantly progressing, it is essential to stay ahead of the curve in protecting your digital assets and personal information.

CRITICAL INFRASTRUCTURE VULNERABILITIES

A key aspect of managing and securing your digital life is understanding critical infrastructure's vulnerability to cyberattacks. This chapter will address its significance, its susceptibility, and what steps can be taken to improve its security and resilience.

Determining Critical Infrastructure

Critical infrastructure refers to physical and digital systems and assets that are vital for our society and economy's smooth operation. These sectors include, but are not limited to:

Energy (power generation, transmission and distribution), Water and wastewater systems

Transportation (aviation, maritime, rail or road), Healthcare & public health Telecommunications and information technology Financial services Agriculture food & agriculture Emergency services Government facilities

Disruption or destruction to these systems could have disastrous effects on public safety, national security, and the economy.

Cyber Vulnerabilities in Critical Infrastructure

Critical infrastructure is becoming more interconnected and dependent on digital technology, increasing the potential for cyber attacks. Cyberattacks on critical infrastructure can cause physical harm, disrupt essential services, or compromise sensitive data. A variety of factors contribute to this vulnerability:

- Legacy Systems: Many infrastructure systems are outdated, leaving them vulnerable to cyberattacks that take advantage of unpatched vulnerabilities.
- Interconnectivity: The increasing interconnection between various infrastructure sectors increases the potential risk of

a cascading effect, in which an attack on one sector could ripple effects throughout others.

- Inadequate cybersecurity measures: Limited resources, lack of awareness and ineffective cybersecurity policies can leave critical infrastructure vulnerable to emerging threats.
- Insider Threats: Disgruntled employees, contractors or other insiders with access to sensitive systems can cause irreparable harm.
- Supply chain risks: Vulnerabilities in components, software or services from third-party suppliers could create holes in critical infrastructure systems.

Enhancing Critical Infrastructure Security

To protect critical infrastructure from cyber threats, we need to take a comprehensive approach that involves collaboration between governments, industry, and individuals. The following measures can help boost the security and resilience of critical infrastructure:

- Create and implement robust cybersecurity policies: Establishing clear guidelines, risk management strategies, and incident response plans can assist organizations in recognizing and mitigating vulnerabilities.
- Conduct Regular Assessment and Upgrading Systems: Regularly audit and upgrade technology, software, and firmware to protect against known vulnerabilities as well as emerging risks.
- Secure the Supply Chain: Implement stringent security measures to guarantee the integrity of third-party components, software, and services.
- Foster a Culture of Cybersecurity Awareness: Educate employees on cybersecurity best practices and the significance of protecting critical infrastructure.
- Enhance Information Sharing: Foster collaboration among government agencies, the private sector, and international partners to share threat intelligence and best practices.

- Invest in Research & Development: Encourage innovation in cybersecurity technology and methods to stay ahead of evolving threats.
- Encourage Personal Responsibility: Individuals can contribute to critical infrastructure security by practicing good cyber hygiene and reporting suspicious activities to relevant authorities.

By understanding the vulnerability of critical infrastructure and taking proactive steps to strengthen its security, we can work together to protect our society's most essential systems and services. As digital technology continues to advance rapidly, it is essential for us to remain vigilant and adaptive in the face of new challenges and threats.

THE POTENTIAL NEGATIVE REPERCUSSIONS OF WIDESPREAD JOB DISPLACEMENT DUE TO AUTOMATION

One significant consequence of our rapidly digitized world is the potential for widespread job displacement due to automation. This chapter will address automation's rise, its effect on workers, and how individuals and society can adjust in light of this shifting landscape.

Automating Manufacturing Operations Automation Is on the Rise

Automation is the use of technology, such as artificial intelligence (AI) and robotics, to perform tasks that were previously handled by humans. Advances in machine learning, computer vision, natural language processing and other fields have hastened automation across various industries such as manufacturing, transportation, retailing and customer service.

Automation offers the potential to enhance efficiency, cut costs and boost productivity; however it also raises concerns about job displacement as machines and algorithms replace human labor.

Job Displacement and Its Effects

The impact of automation on employment will likely be uneven, with some jobs more vulnerable than others. Generally, tasks that are repetitive, routine or predictable are at higher risk for automation; conversely, jobs requiring creativity, problem-solving abilities, emotional intelligence and interpersonal abilities have fewer chances for automation.

Automation could lead to widespread job displacement, potentially having several detrimental effects:

- Unemployment: A dramatic spike in joblessness, especially among low-skilled workers and those employed in occupations particularly susceptible to automation.
- Economic Inequality: Automation could disproportionately impact low-wage workers, widening income disparities and deepening social stratification.
- Social Unrest: High unemployment and rising inequality could create social and political unrest as affected individuals struggle to adjust in a rapidly evolving job market.
- Loss of Consumer Purchasing Power: With fewer people employed, consumer spending may decrease, potentially impeding economic growth.

Adopting to the Age of Automation

To prevent job displacement from having an adverse impact, individuals and society must take proactive measures to prepare for the age of automation:

- Adopt a lifelong learning mindset: Stay ahead of the job market by consistently improving your skills and knowledge base. This could include studying new technologies, earning certifications, or pursuing higher education.
- Focus on transferrable skills: Hone skills that can be applied across numerous industries and occupations, such as critical thinking, communication, teamwork, and adaptability.
- Pursue Careers in Resilient Industries: Look into fields that are less susceptible to automation, such as healthcare, education and creative industries.
- Support Policies That Foster Job Creation: Advocate for policies that stimulate the development of new industries, provide training programs for workers displaced by automation, and offer financial assistance to those displaced by these changes.

- Encourage Entrepreneurship: Encourage innovation and support entrepreneurs who create new businesses, products, and services that could create employment opportunities.
- Investment in Social Safety Nets: Enhance social welfare programs like unemployment insurance, retraining initiatives, and income support to aid workers as they transition into new careers or industries.
- Promoting the Benefits of Automation: While it's necessary to address potential job losses, it is also essential to recognize its positive aspects, such as increased productivity, enhanced safety and the potential creation of new, higher-quality jobs.

By anticipating the potential effects of automation on job displacement and taking proactive steps to adapt, individuals and society can better prepare for the coming challenges of the digital age and build a more resilient future. With thoughtful consideration, strategic planning, and an enthusiasm for lifelong learning, we can seize the advantages offered by automation while mitigating its negative effects.

LOSS OF PRIVACY AND THE ABOLISHING OF CIVIL LIBERTIES

As technology develops at an incredible rate, so too does the potential for surveillance and intrusion into our lives. In this chapter, we'll address how privacy has been lost and civil liberties diminished in today's digital age. We'll consider both the consequences of these changes as well as strategies you can employ to protect your privacy and rights as a prepper.

The Digital Landscape

The digital age has brought many conveniences, but it also created a data-driven world where personal information is valued as currency.

With every online purchase, social media post or app download we leave behind an electronic trail which can be tracked, analyzed and exploited. In turn this data informs our digital experiences - from personalized advertisements to carefully curated news feeds - for better or worse.

Government Surveillance

The erosion of privacy isn't just limited to corporations. Governments around the world are increasingly using technology for citizen monitoring, from warrantless wiretapping and mass data collection, to facial recognition in public places. While these measures may be meant to safeguard national security, they can also be employed against dissent and suppress civil liberties.

Corporate Intrusion

Corporations' interests are driving the erosion of privacy as they collect vast amounts of consumer data to build profiles and target advertising. This gives corporations unprecedented control over individual behavior through algorithms that predict our choices. Furthermore, data breaches and the sale of personal information to third parties further compromise privacy and security.

Social Media and Privacy

Social media platforms have become a vital part of modern life, but they also pose an immense privacy risk. By nature, these sites encourage users to share personal information and build extensive networks of connections - data which could then be harvested for illicit purposes such as identity theft or targeted harassment.

The Internet of Things (IoT)

As more devices become interconnected, the potential for privacy breaches increases. IoT devices such as smart speakers, security cameras and wearable technology can collect and transmit sensitive data which leaves users vulnerable to hacking or surveillance attempts. Furthermore, these IoT gadgets could be commandeered for cyberattacks that could compromise entire networks' security measures.

Biometrics and Surveillance

Biometric technology, such as facial recognition and fingerprint scanners, has become more widely used in recent years. While these advances offer convenience and security, they also raise concerns about abuse potentials.

The collection and storage of biometric data creates opportunities for privacy breaches and the erosion of civil liberties since individuals can be tracked, profiled, and targeted based on their unique physical characteristics.

Digital Currency and Financial Privacy

Digital currencies and electronic payment systems offer the promise of efficient, frictionless transactions. Unfortunately, they also present risks to financial privacy as transactions are typically recorded on publicly accessible ledgers. This allows for tracking individual spending habits as well as government surveillance or censorship initiatives.

Strategies for Privacy Preservation

In light of these obstacles, preppers can take measures to safeguard their privacy and civil liberties. These may include:

- Utilizing encrypted communication tools, such as secure messaging apps and email encryption.
- Utilizing virtual private networks (VPNs) to mask browsing activity and location data.
- Regularly reviewing and updating privacy settings on social media platforms.
- Exercise caution when sharing personal information online and avoid oversharing on social media platforms.
- Use strong, unique passwords and two-factor authentication for all online accounts to maintain the security of those accounts.
- Education on the risks and advantages of IoT devices, and taking steps to secure them on home networks.
- Supporting privacy-enhancing technologies such as end-to-end encrypted communication platforms and decentralized digital currencies.
- Advocating for privacy legislation and the protection of civil liberties at local, national, and international levels.
- Staying abreast of developments in surveillance technology and privacy issues so that you can make informed decisions about which tools and platforms you utilize.

In today's digital world, privacy and civil liberties are becoming more and more at risk. As preppers, it's up to us to stay vigilant and adapt. By understanding the risks, staying informed, and taking proactive steps to protect your privacy online, you can better prepare yourself for life's unexpected surprises in a rapidly-evolving landscape.

PART V

PREPARING FOR A DIGITAL ARMAGEDDON

CONSTRUCTING A DIGITAL BUG-OUT BAG

In today's digitally connected world, it is essential to have a digital strategy as part of your overall preparedness plan. A digital bug-out bag (DBOB) is similar to the physical bug-out bag preppers carry during emergencies. This chapter will guide you through creating your digital bug-out bag so that you have access to essential information, tools and services during times of emergency.

Secure Digital Storage

The initial step in creating a DBOB is selecting an effective secure digital storage solution. Ideally, select a device that's portable, reliable and has plenty of capacity for your files. Popular options include:

- USB Flash Drives: These compact devices are ideal for storing important documents and files. Opt for one with at least 64GB of storage capacity and a durable design to withstand harsh conditions.
- External Hard Drives: For larger storage requirements, an external hard drive offers more capacity than a flash drive. Make sure to choose a ruggedized, waterproof and shockproof model.
- Cloud storage: Services like Dropbox, Google Drive and iCloud provide a remote storage solution that is accessible from any device with an internet connection. Be sure to create a strong password for added security and enable two-factor authentication for added protection.

Important Documents

Gather copies of essential documents and organize them into folders on your storage device. Some essential documents to include:

- Identification: Birth certificates, passports, driver's licenses and Social Security cards.
- Medical: Immunization records, prescription information and copies of health insurance cards.
- Financial: Bank account/credit card info, tax returns and a list of financial assets and liabilities.
- Legal: Wills, power of attorney documents and any other legal paperwork.
- Insurance: Copies of home, auto, life and other insurance policies.
- Contacts: A list of important phone numbers and email addresses for family members, friends and emergency contacts.

Digital Tools

Your DBOB should contain essential digital tools and software that can assist in times of emergency. These may include:

- Maps: Download offline maps of your local area and possible evacuation routes. Google Maps allows for saving maps for offline use.
- Communication Apps: Install secure messaging applications like Signal or WhatsApp which provide end-to-end encryption, guaranteeing conversations remain private.
- Emergency Apps: Download apps that provide real-time updates on weather conditions, natural disasters, and other emergency situations. Examples include FEMA, American Red Cross, and Weather Underground.
- Backup Power: Make sure you have portable chargers or solar-powered chargers on hand in case of emergency, so your devices remain charged.

Encryption and Security

Protecting your sensitive information is paramount. Utilize encryption software like VeraCrypt or BitLocker to create encrypted containers for important files, so even if the storage device gets stolen, your data remains secure.

Regular Updates

Your digital bug-out bag requires periodic upkeep just like your physical bug-out bag. Make sure all documents and files are updated at least once annually or whenever there are major life changes such as moving or changing jobs. Furthermore, be sure to check for software updates for all digital tools to guarantee they remain functional and secure.

Multiple Copies and Accessibility

Finally, create multiple copies of your DBOB and store them in various places such as at home, in your car, and with a trusted friend or family member. Doing this ensures that you can access essential information even if one copy is lost or damaged.

By following these steps, you'll create a comprehensive and secure digital bug-out bag (DBOB) that will serve you well during times of emergency. Remember that preparedness in the digital age is just as important as physical preparedness; with an effective DBOB, you'll have peace of mind knowing you have all necessary information, tools, and resources to navigate emergencies in our increasingly digital world. To keep it effective and up to date, be sure to regularly update and maintain it just like with physical bug-out bags. Staying diligently and proactive about preparing ahead will give yourself better equip ourselves to tackle whatever obstacles may come our way in the future.

ENHANCING PERSONAL CYBERSECURITY

In today's ever-connected world, personal cybersecurity is of utmost importance. Protecting your sensitive information is essential for maintaining safety and privacy online. This chapter will give you essential strategies and tips to enhance your personal protection and reduce the likelihood of becoming a victim of cyberattacks.

Strengthen Your Passwords

Weak passwords are one of the most common ways cybercriminals gain unauthorized access to accounts. To bolster your cybersecurity:

- Create strong, unique passwords for each account using a combination of uppercase and lowercase letters, numbers, and special characters.
- Steer clear of easily guessable information like names, birthdays, or common words.
- Consider using a passphrase--a series of random words or memorable sentence that is harder to crack.
- Additionally, utilize trusted password managers like LastPass or Dashlane for assistance in creating and storing secure passwords.

Enable Two-Factor Authentication (2FA)

Two-factor authentication (2FA) adds an extra layer of protection by requiring you to verify your identity using a secondary method, such as text message, authenticator app, or physical token. Enable 2FA on all accounts that offer it - especially those containing sensitive information like email, financial accounts, and social media accounts.

Keep Software and Devices Up-to-Date

Outdated software and operating systems can leave you vulnerable to cyberattacks. Regularly upgrade your devices, applications, and antivirus software so that you're protected against known vulnerabilities and threats. When possible, enable automatic updates for added convenience.

Be Wary of Phishing Attempts

Phishing attacks use emails, texts or social media messages that appear legitimate from legitimate sources in an effort to trick you into providing sensitive information or downloading malware. In order to avoid being duped by these scams:

- Be wary of unsolicited messages, particularly those that request personal information or require you to take immediate action.
- Verify the sender's email address and look out for any suspicious elements such as poor grammar or spelling mistakes.
- Hover over links without clicking to verify if they lead to legitimate websites.
- Utilize antivirus software with built-in phishing protection.

Secure Your Home Network

Your home network can be an entry point for cybercriminals; in order to protect both yourself and your devices from this risk, be sure to take these precautions:

- Change the default login credentials on your router and use a strong, unique password.
- Enable WPA3 encryption - the most secure Wi-Fi protocol available.
- Regularly upgrade your router's firmware.

- Disable remote management and guest network features unless absolutely necessary.
- Consider using a virtual private network (VPN) to encrypt your internet connection.

Be Wary When Utilizing Public Wi-Fi

Public Wi-Fi networks tend to be unprotected, making them prime targets for hackers. When using public Wi-Fi:

- Avoid accessing sensitive accounts, like email or financial services.
- Encrypt your connection and use a VPN to encrypt it from prying eyes.
- Disable file sharing and use a firewall to block unauthorized access to your device.

Safeguard Your Social Media Presence

Establish controls over what content you post online using various tools like firewalls.

Cybercriminals may harvest personal information from your social media profiles to launch targeted attacks against you. To protect yourself:

- Limit the amount of personal information you share online.
- Adjust your privacy settings to restrict who can view your content.
- Exercise caution when accepting friend requests and connecting with strangers.
- Regularly review and update social media security settings as necessary.

Back Up Your Data

In the event of a cyberattack, having a backup of your critical files can help you recover more quickly. Store them on an external hard drive or secure cloud storage service - or both! Encrypt and password protect these backups for extra protection.

By following these strategies, you can significantly strengthen your personal cybersecurity and reduce the likelihood of falling victim to cyberattacks. Maintaining personal cybersecurity requires ongoing maintenance and vigilance; stay abreast of new threats and best practices so you remain protected in today's ever-evolving digital landscape. As a prepper in today's digital age, commitment to cyber-security is not only necessary for safety and privacy but also accessing vital information and resources during times of crisis.

BUILDING SELF-RELIANCE SKILLS

The Digital Age has provided us with an abundance of tools and resources that can make our lives simpler and more connected. However, relying solely on technology can leave us vulnerable when systems malfunction or become compromised. As a prepper, developing self-reliance skills is essential to ensure you can handle both digital and physical challenges. This chapter will guide you through the essential self-reliance skills to master in the Digital Age.

Learning Basic First Aid

In times of emergency, access to medical care may be restricted or delayed. Be familiar with basic first aid techniques like CPR, treating wounds, stabilizing fractures and recognizing common medical signs. Consider taking a certified first aid course for further practice and confidence.

Gain Communication Skills

Knowing alternative communication methods when digital channels are unavailable can be invaluable. Get familiar with:

- Two-Way Radios: Get yourself a set of two-way radios and become familiar with their operation and range limitations. Learn the phonetic alphabet and standard radio procedures so that you can communicate effectively.
- Signaling: Acquire basic signaling techniques, such as using a whistle, signal mirror, or flares to alert others of an emergency.
- Non-digital Communication: Master the art of crafting concise messages and developing effective listening skills for face-to-face interactions.

Master Analog Navigation

GPS devices and smartphones offer convenient navigation, but they may malfunction or lose signal. Increase your self-reliance by learning traditional navigation methods:

- Map Reading: Acquire the skills to read topographic maps and identify features like terrain, elevation, and water sources.
- Compass Skills: Get acquainted with using a compass and taking bearings while navigating with a map in hand.
- Natural Navigation: Learn to navigate using the sun, stars and other natural indicators.

Acquire Basic Survival Skills

Equip yourself with the necessary survival knowledge so that you can manage in times of emergency. Some essential skills include:

- Fire-starting: Discover various techniques for starting a fire, such as using a fire starter, using flint and steel or friction-based methods.
- Shelter Building: Master the art of creating temporary shelters from materials found in nature or items from your emergency kit.
- Water Purification: Be familiar with various water purification methods, such as boiling, chemical treatment and using portable filters.

You can learn the above essential survival skills and other in 'Evasive Wilderness Survival Techniques'.

https://www.sfnonfictionbooks.com/evasive-wilderness-survival-techniques

Grow and Preserve Food

Growing your own food provides you with self-sufficiency and independence from digital technology. Learn how to:

- Garden: Gain proficiency in gardening basics such as planting, soil management and pest control.
- Foraging: Recognize edible plants and fungi native to your region.
- Preservation: Learn food preservation techniques like canning, drying and fermenting.

Develop Basic Repair & Maintenance Skills

In times of emergency, you may need to rely on your ability to repair and maintain essential items. Develop skills such as:

- Basic carpentry: Discover how to utilize hand tools to construct and repair basic structures.
- Sewing: Master fundamental sewing techniques in order to repair clothing as well as create useful items like bags and pouches.
- Mechanical Repair: Gain an in-depth knowledge of how to diagnose and repair basic mechanical issues with vehicles, generators, and other equipment.

By honing these self-reliance skills, you'll be better equipped to meet any challenges the Digital Age may present. Remember, self-reliance is a lifelong journey and it's essential to practice and hone your abilities on a regular basis. As a prepper in the Digital Age, balancing digital preparedness with self-reliance techniques will give you the skillset needed to tackle any situation that comes your way.

PART VI

NAVIGATING THE PROMISE OF A TECHNOLOGICAL UTOPIA

ADOPTING LIFELONG LEARNING AND ADAPTABILITY

In today's fast-paced world, lifelong learning and adaptability are essential for survival - especially digital preppers. The digital revolution has drastically changed how we live, work, and communicate, offering both new challenges and opportunities. As a prepper, your success depends on your capacity for adapting in this ever-evolving landscape. In this chapter we'll discuss the significance of lifelong learning as well as provide practical strategies to develop adaptability skills.

The Importance of Lifelong Learning

The digital age is marked by rapid advances in technology, making it essential for preppers to stay abreast of their skills and knowledge. A commitment to lifelong learning ensures you remain relevant and resilient in the face of digital uncertainties. Here are a few reasons why lifelong learning is essential for digital preppers:

- Staying abreast of technological advancements: New tools and techniques are being created at an incredible speed, so staying informed allows you to take advantage of these advances for enhanced preparedness.
- Enhancing Problem-Solving Skills: As you gain knowledge, you will become better prepared to handle unexpected circumstances and think creatively.
- Extending Your Network: Lifelong learning offers you the chance to connect with like-minded individuals, strengthening your support system and access to resources.

Strategies for Lifelong Learning

Now that we have established the importance of lifelong learning, let us look at some practical strategies to help you embrace it:

Establish Specific Learning Objectives: Identify areas where you need to develop or expand your knowledge, and set achievable yet attainable objectives for the duration of your educational journey.

- Engage in Online Learning: Take advantage of the wealth of resources available online, such as webinars, podcasts and e-books, to develop your skills and insight.
- Participate in workshops and seminars: Attend events that target the skillsets you wish to acquire, not only will this enhance your understanding but also allow you to expand your network.
- Connect to Meaningful Communities: Find like-minded individuals through online forums, social media groups or local meet-ups. These gatherings offer support, inspiration and invaluable insights.
- Teach Others: Spread your knowledge with others to deepen your understanding and assist fellow preppers on their journey.

Fostering Adaptability Through Cultivation

Adaptability is the capacity to adjust to new conditions or circumstances. In today's digital world, change is constant and cultivating adaptability is essential for digital preppers. Here are a few strategies to help you become more adaptable:

- Create a growth mindset: Accept challenges and see them as chances for learning and improvement. This attitude will give you the courage to tackle new situations with assurance.
- Practice flexibility: Be open to new ideas and willing to adjust your plans as more information becomes available. Flexibility can help you effectively address unexpected obstacles.
- Stay Curious: Foster a curious mindset by asking questions, exploring new ideas, and seeking out new experiences.

Curiosity fosters adaptability by keeping you engaged and open to change.

- Manage Stress Effectively: Develop healthy coping mechanisms to manage stress, as it can hinder your ability to adapt. Techniques such as mindfulness, exercise and journaling can help reduce tension and promote mental clarity.
- Learn from Others: Pay attention to how others manage change and adapt in different circumstances. You can gain valuable lessons from their successes as well as their failures.

By adopting a lifelong learning and adaptability mindset, you will be better equipped to deal with the unknowns of the digital age. As a digital prepper, your success is directly tied to your capacity for knowledge acquisition, adaptation, and growth. Invest in yourself and your future by dedicating to continuous education and cultivating an adaptable mindset - this will guarantee you remain resilient and prepared no matter what challenges come your way in this brave new world!

ESTABLISHING A HEALTHY RELATIONSHIP WITH TECHNOLOGY

In today's digital age, technology has become an integral part of our lives. While it provides us with many amazing advantages, it can also present challenges when trying to maintain a healthy balance. As prepper, it's essential that we develop an appropriate relationship with technology so that it serves as an asset rather than hinderance. In this chapter, we'll examine why having a balanced approach towards technology is so important and provide practical tips for cultivating one.

The Need for a Balanced Approach

A balanced approach to technology ensures you reap its rewards without becoming overly dependent or consumed by it. Here are a few reasons why this type of approach is essential for digital preppers:

- Enhancing Preparedness: Technology can offer invaluable tools and resources to increase your preparedness.
  However, an over-reliance on it could leave you vulnerable when it malfunctions or becomes unavailable.
- Reducing Digital Fatigue: Constant connectivity can lead to digital fatigue, which has detrimental effects on mental and physical wellbeing. A balanced approach helps you maintain your health while staying focused on prepping objectives.
- Protecting Privacy and Security: Maintaining a healthy relationship with technology involves being aware of potential threats and taking measures to protect yourself.

Tips for Fostering a Healthy Relationship with Technology

Here are a few tips that can help you foster an harmonious relationship with technology and maintain a balanced approach:

- Establish Clear Boundaries: Set boundaries on your technology use, such as designated times for checking email or engaging on social media. Doing this will help prevent you from becoming consumed by digital distractions.
- Prioritize Offline Skills: While technology can be an invaluable asset, it's essential to continue honing and honing offline abilities as well. Make time to practice traditional prepping techniques like gardening, first aid and navigation so that you remain self-reliant in various situations.
- Practice Digital Detoxifications: Take regular breaks from technology for short periods of time to recharge and refocus. This could be as simple as going for a daily walk without your phone or taking a weekend camping trip with limited digital access.
- Be selective in the technology you select: Not all technology is created equal. Focus on tools and platforms that directly support your prepping goals and enhance preparedness; avoid those which act as mere distractions or offer minimal value.
- Stay Aware of Potential Hazards: Stay informed on the newest cybersecurity threats and privacy issues. This knowledge can help you make informed decisions about which technology to use and what precautions to take.
- Implement Security Best Practices: Protect your digital assets by employing strong passwords, using encryption, and keeping software up-to-date. Regularly back up data and consider investing in secure offline storage solutions for essential information.
- Foster Healthy Tech Habits: Refrain from using technology as a crutch or escape from reality. Instead, utilize it as an

aid to assist with prepping goals and enhance your overall well-being.

By cultivating a healthy relationship with technology, you can reap its rewards without becoming overly dependent or endangering yourself. As a digital prepper, it's essential to find a balance that allows technology to serve as an asset in your preparedness journey. With thoughtful consideration and intentional habits, maintaining this balanced relationship with technology supports your objectives and increases resilience in today's digital age.

BALANCING TECHNOLOGY WITH HUMAN INTERACTION

As digital preppers, it's essential to recognize that technology can be beneficial but cannot replace human connection. Human connections are fundamental for our well-being, mental health and overall preparedness. In this chapter we'll examine the significance of human interaction and offer strategies for striking a balance between technology use and cultivating meaningful relationships.

The Importance of Human Interaction

Human interaction is essential for many reasons, such as:

- Emotional Support: Emotional connections and support systems are essential during times of hardship. People often come together in times of need and rely on one another for comfort, guidance, and assistance.
- Skill-sharing and Knowledge Exchange: Engaging with others allows you to exchange skills, insights, and experiences that can strengthen your preparedness. Face-to-face communication offers opportunities for hands-on learning as well as more efficient knowledge transfer.
- Strengthening Community Ties: Fostering strong connections within your community can be invaluable during times of crisis, when people come together to offer mutual assistance and support.
- Enhancing Decision-Making: Human interaction encourages collaboration and diverse viewpoints, which can result in improved decision-making and problem solving abilities.

Strategies for Harmonizing Technology and Human Interaction

To achieve a balance between technology and human connections, try these strategies:

- Establish Boundaries for Technology Use: Set limits on your technology use, especially during social gatherings or family time. Doing this will help ensure that digital devices do not hinder interactions with others.
- Utilize Technology for Connections: Take advantage of technology by using it to foster relationships. Video calls, messaging apps, and social media can help you stay in touch with friends and family even when physical distances are an issue.
- Prioritize face-to-face interactions: Wherever possible, prioritize in-person communication over digital alternatives. Face-to-face encounters can create deeper connections and enable more efficient communication.
- Engage in Group Activities and Skill Sharing: Arrange or attend group events like workshops, community projects, or skill sharing sessions to foster connections while honing your preparedness skills. These activities can help you form new connections while building upon existing ones.
- Cultivate a Support Network: Establish and nurture an extensive support system of family, friends, and fellow preppers. This group can offer invaluable resources, emotional comfort, and practical help when necessary.
- Join local organizations and groups: Joining community organizations, clubs, or prepping groups is an excellent way to build connections and collaborate on shared objectives. These gatherings can offer a wealth of knowledge, resources, and hands-on learning experiences.
- Maintain a Healthy Balance: Evaluate your technology use and its effect on relationships regularly. Be mindful of potential imbalances, then adjust your habits to achieve an

equilibrium between digital technology use and human interaction.

By striking a balance between technology and human connection, you can ensure your preparedness efforts are comprehensive and successful. Human connections are essential for resilience and adaptability; nurturing these relationships will give you invaluable tools to tackle the obstacles of the digital age. As a digital prepper, it's essential to recognize the significance of human contact and make conscious efforts to maintain healthy balance in your life.

HARNESSING TECHNOLOGY FOR PERSONAL AND COMMUNITY RESILIENCE

As we navigate the digital age, technology plays an increasingly significant role in increasing personal and community resilience. By using it effectively, digital preppers can improve their preparedness efforts, build support networks, and contribute to the overall resilience of their communities. In this chapter, we'll look at practical strategies for utilizing technology for resilience building purposes and discuss its potential uses.

Enhancing Individual Resilience through Technology

Technology can greatly enhance personal resilience in the following ways:

- Access to Information: The internet provides an abundance of knowledge on prepping topics, from emergency preparedness to sustainable living. By tapping into these resources, you can build the knowledge and skillset necessary for becoming more self-reliant.
- Online Learning Opportunities: Online courses, webinars, and tutorials offer invaluable chances to hone your skills and knowledge in a range of topics from first aid to gardening.
- Digital Tools for Organization and Planning: Utilize digital tools, such as apps and software, to efficiently plan, organize, and monitor your prepping efforts - from inventory management to emergency contact lists and disaster response plans.
- Communication and Networking: Technology provides you with the ability to connect with like-minded individuals, exchange ideas, and collaborate on projects. These connections can help you cultivate a powerful support system and access invaluable resources.

- Early Warning Systems and Alerts: Stay informed about potential threats and disasters by subscribing to alert systems like weather apps or emergency notifications from local authorities.

Enhancing Community Resilience with Technology

Technology can also be utilized to promote community resilience in several ways:

- Strengthening Community Connections: Utilize social media platforms and online forums to engage your community, exchange information, and collaborate on projects that benefit everyone involved.
- Coordinated Emergency Response: Technology can facilitate efficient communication and coordination during crises. For instance, community members can utilize messaging apps to share details about available resources, road closures, or evacuation routes.
- Crowdsourcing Solutions: Utilize online platforms to solicit ideas and collect input from community members on various topics, such as local sustainability initiatives or disaster preparedness plans.
- Mapping and Data Visualization: Utilize digital tools to map out and visualize community resources, hazards, and vulnerabilities. This data is invaluable for planning and decision-making at the community level.
- Education and Skill-Sharing: Host or participate in online workshops, webinars, and skill-sharing sessions to foster knowledge exchange and skill development within your community.

Strategies for Leveraging Technology for Resilience

To maximize technology's potential for personal and community resilience, consider these strategies:

- Prioritize Relevant Technologies: Focus on technologies that directly contribute to your preparedness objectives and enhance personal and community resilience.
- Stay Ahead of Technological Advancements: Stay informed on the newest technological innovations and assess their potential benefits for personal and community resilience efforts.
- Implement Security Best Practices: Safeguard your digital assets and privacy by using strong passwords, using encryption, and keeping software up-to-date.
- Educate Others: Spread your understanding of technology and its benefits for resilience with others in your community. Doing so can help create a culture of preparedness and collective action in your area.
- Collaborate and innovate: Partner with others in your community to explore innovative technologies for resilience-building initiatives and projects.

By effectively using technology, digital preppers can significantly enhance their personal and community resilience. Through effective use of this powerful tool, you can improve preparedness efforts, strengthen support networks, and contribute to the overall resilience of your community. As a digital prepper it's essential to recognize its value and make conscious efforts to use it in ways that promote personal and community resilience objectives. Through thoughtful consideration and intentional collaboration, you can take advantage of technology's potential to create a more resilient future for yourself and those around you in this digital age.

CONCLUSION

In "The Prepper's Guide to the Digital Age," we have explored many aspects of technology and its effects on our lives, from understanding its landscape to assessing its potential advantages and risks. Throughout the book, we emphasize the significance of preparedness, adaptability, and resilience as we navigate an increasingly interconnected and technologically advanced world.

At the end of our journey, it is essential to remember that technology can be a double-edged sword. While it offers us incredible opportunities for growth and knowledge, it also poses risks to privacy, security and even mental health. As digital preppers we must strive to strike a balance between utilizing its benefits while mitigating any potential downsides.

The key takeaway from this guide is the necessity of lifelong learning and adaptability in our rapidly advancing world. By actively seeking new knowledge, skills, and experiences, we can better prepare ourselves for both the opportunities and challenges the digital age presents. Additionally, maintaining a healthy relationship with technology while balancing its use with meaningful human interactions will guarantee our wellbeing while cultivating social connections which are so crucial for resilience.

In today's digital age, it is critical that we foster a sense of community and work together to build resilience on both an individual and collective level. Through technology-enabled personal and community resilience initiatives, we can improve our preparedness efforts, strengthen support networks, and contribute to society's overall wellbeing.

The future is uncertain and the digital age presents us with a myriad of challenges and opportunities. As we continue to navigate this turbulent new world, let us embrace principles like preparedness, adaptability, and resilience - using technology as an enabler in our quest for self-reliance and community strength.

May this guide serve as a springboard for your digital prepping journey, providing you with the skillset and strategies needed to succeed in today's digital era. Let us face the future together, equipped with wisdom and determination in order to build a more resilient, prepared, and connected world.

THANKS FOR READING

Dear reader,

Thank you for reading *The Prepper's Guide to the Digital Age.*

If you enjoyed this book, please leave a review where you bought it. It helps more than most people think.

Don't forget your FREE book!

You will also be among the first to know of FREE review copies, discount offers, bonus content, and more.

Go to:

www.SFNonfictionBooks.com/Free-Book

Thanks again for your support.

AUTHOR RECOMMENDATIONS

Teach Yourself Escape and Evasion Tactics!

Discover the skills you need to evade and escape capture, because you never know when they will save your life.

Get it Free Today

www.SFNonfictionBooks.com/Free-Book

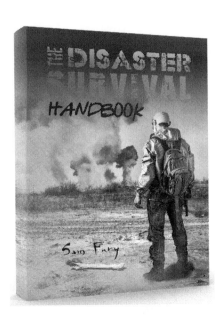

Do You Know How to Survive a Disaster?

This book is a must-have in your disaster survival kit, because the information will save your life.

Get it Free Today

ABOUT SAM FURY

Sam Fury has had a passion for survival, evasion, resistance, and escape (SERE) training since he was a young boy growing up in Australia.

This led him to years of training and career experience in related subjects, including martial arts, military training, survival skills, outdoor sports, and sustainable living.

These days, Sam spends his time refining existing skills, gaining new skills, and sharing what he learns via the Survival Fitness Plan website.

www.SurvivalFitnessPlan.com

Made in the USA
Columbia, SC
30 August 2023

22297846R10057